FAMOUS BRIDGES OF THE WORLD

Measuring Length, Weight, and Volume

Yolonda Maxwell

PowerMath™

The Rosen Publishing Group's
PowerKids Press™
New York

Published in 2005 by The Rosen Publishing Group, Inc.
29 East 21st Street, New York, NY 10010

Book Design: Michael Tsanis

Photo Credits: Cover © Stephen Simpson/Taxi; p. 5 © Michael S. Yamashita/Corbis; p. 6 © Dallas and
John Heaton/Corbis; p. 9 © Tom Bean/Corbis; p. 10 © Robert Estall/Corbis; p. 13 © James Davis;
Eye Ubiquitous/Corbis; pp. 14–15 © Will and Deni McIntyre/Corbis; p. 17 © Angelo Hornak/Corbis;
p. 19 © Alan Schein Photography/Corbis; p. 23 © Galen Rowell/Corbis; p. 25 © Robert Essel NYC/Corbis;
p. 27 © Bettman/Corbis; p. 28 © Raymond Gehman/Corbis; p. 30 © Jose Fuste Raga/Corbis.

Library of Congress Cataloging-in-Publication Data

Maxwell, Yolonda.
  Famous bridges of the world : measuring length, weight, and volume / Yolonda Maxwell.
      p. cm. — (PowerMath)
  Includes index.
  ISBN 1-4042-2937-X (library binding)
  ISBN 1-4042-5137-5 (pbk.)
  6-pack ISBN 1-4042-5139-1
  1. Mensuration—Juvenile literature. 2. Bridges—Juvenile literature. I. Title. II. Series.
  QA465.M39 2005
  516'.15—dc22
                              2004007065

Manufactured in the United States of America

Contents

Engineering Marvels

Bridges are such a familiar part of life today that we often take them for granted. Have you ever considered what challenges **engineers** face when building these structural marvels, or how our lives would be different without them? To get an idea of how bridges affect your life, imagine a world where there are no bridges.

Think about a bridge you cross often. Perhaps you cross one on your way to school or to visit family or friends in other places. Without this bridge, how would you get where you needed to go, and how long would it take you to get there? (Don't imagine that you'd use another nearby bridge, because there are no bridges anywhere.)

If you wanted to build a bridge, how would you go about it? What would you need to know? Engineers need to know the properties of the building materials they use. They also need math skills to calculate the quantities of building materials they'll need. We can use our math skills to learn about the lengths, weights, and volumes of materials that played a part in the construction of some of the world's most famous bridges.

The earliest bridges were simple, sometimes no more than logs laid across streams or gorges, or ropes stretched across rivers or valleys. This photograph, taken in the 1980s, shows people using an ancient rope bridge in southern China.

weight

Arch Bridges

When engineers design bridges, they must make sure the bridge is strong enough to support its own weight plus the weight of the traffic it will carry. Many early bridge builders favored the arch bridge because of its great strength. Instead of pushing straight down on the bridge's **deck**, the weight of both traffic and the bridge is transferred outward along the curves of the arch to the large supports at each end. These supports, called **abutments**, carry the weight and keep the ends of the arch from spreading out.

Ancient Romans, who are still known today for their engineering feats, built the structure shown on page 6 around 2,000 years ago. Located in what is today southern France, the stone structure is both a bridge and an **aqueduct**. This section over the Gard (GAHR) River is known by its French name, Pont du Gard, which means "Bridge over the Gard."

The Pont du Gard has 3 levels, each with a series of arches. A channel on the top level carried water to the city of Nîmes (NEEM). A road ran along the top of the lowest level.

The Pont du Gard is what remains of a structure that was originally much longer. The ancient aqueduct stretched from Nîmes to a water source about 30 miles away.

The Gard River valley is wider at the top than at the bottom, so the middle level of the Pont du Gard is longer than the bottom level, and the top level is longer than the middle level. The bottom level is about 510 feet long and has 6 large arches. The middle level is about 870 feet long and has 11 large arches. The top level has 35 small arches, each about 16 feet wide. Each of the 36 pillars on the top level is about 9.5 feet wide. What is the total length of the top level?

First, multiply 16 feet by 35 to find the length of all the arches.

```
   16  feet per arch
 x 35  arches
   80
 + 48
  560  feet
```

Next, multiply 9.5 feet by 36 to find the length of all the pillars.

```
    9.5  feet per pillar
 x  3 6  pillars
   57 0
 + 285
  342.0  feet
```

Finally, add these 2 products to find the length of the upper level.

```
   560  feet (length of arches)
 + 342  feet (length of pillars)
   902  feet (total length)
```

The total length of the upper level is about 902 feet.

Altogether, the Pont du Gard is about 160 feet high. It is built of limestone blocks, some of which weigh as much as 4,000 pounds. The blocks were cut so precisely and fit together so tightly that no cement was needed to hold the stones in place.

The aqueduct carried enough water to provide each of the 440,000 people in Nîmes with about 100 gallons of water each day. How many gallons of water did the aqueduct carry to Nîmes every day? To find the answer, multiply the number of people by the number of gallons per person.

Abraham Darby III, the builder of the Iron Bridge, was very proud of his achievement. He had the date of the bridge's construction inscribed on the outer arch, and hired artists to make paintings and engravings of the finished bridge.

Another famous arch bridge is the Iron Bridge, built across the Severn Gorge in Shropshire, England, in 1779. At first, the Iron Bridge may not look very impressive. It is much smaller than the Pont du Gard. The arch **spans** only 100 feet. What makes the bridge famous is that it was the first bridge in the world to be made entirely of **cast iron**.

The Iron Bridge's arch is actually made up of 5 thin arches, called ribs. Each rib was made in 2 parts, which were moved from the factory to the bridge site and attached to each other after they were in place. Each $\frac{1}{2}$ rib weighs about 5.75 tons. How many tons do the 5 ribs weigh altogether?

First, multiply 5.75 tons by 2 to find the weight of a single rib.

$$\begin{array}{r} 5.75 \text{ tons per half rib} \\ \times \quad 2 \text{ half ribs} \\ \hline 11.50 \text{ tons} \end{array}$$

Next, multiply 11.5 tons by 5 to find the total weight of all 5 ribs.

$$\begin{array}{r} 11.5 \text{ tons per rib} \\ \times \quad 5 \text{ ribs} \\ \hline 57.5 \text{ tons} \end{array}$$

The total weight of all 5 ribs is about 57.5 tons.

In 2001, the British Broadcasting Corporation (BBC) built a half-size model of the Iron Bridge to learn more about how it was built and what kind of problems the builders faced. One thing they learned was that it would have taken about 25 men to lift $\frac{1}{2}$ of a single rib!

How many pounds did all 5 ribs weigh? To find the answer, multiply the number of tons (57.5) by the number of pounds per ton (2,000).

Arch bridges worked so well that engineers continued to build them into the twentieth century. One famous twentieth-century arch bridge is the Sydney Harbour Bridge, which was built between 1924 and 1932 to carry traffic across the harbor of Sydney, Australia. The Sydney Harbour Bridge differs from the Pont du Gard and the Iron Bridge in that the arch is above the deck rather than below it. However, the arch still carries the weight of the bridge. Steel hangers suspend the deck from the arch above it.

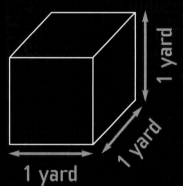

1 yard

When construction began on the bridge, workers first dug 4 deep holes in the bottom of the harbor for the bridge's foundations. The amount of rock removed from each hole was measured in cubic yards. A cubic yard is the volume of a cube whose height, width, and depth are each 1 yard.

Workers removed about 39,955 cubic yards of rock from each hole. About how many cubic yards of rock did they remove altogether? To find the answer, multiply the number of cubic yards removed from each hole (39,955) by the number of holes (4).

39,955 cubic yards per hole
x   4 holes
159,820 cubic yards

Workers removed a total of about 159,820 cubic yards of rock for the foundations of the Sydney Harbour Bridge.

Once the construction was completed, workers gave the bridge 3 coats of paint. If they used about 24,027 gallons of paint for each coat, how many gallons did they use altogether?

Beam Bridges

Another common kind of bridge is a beam bridge. A beam bridge consists of a **horizontal** beam supported by **piers**. This is the simplest and least expensive type of bridge to build. Beam bridges work well when the distance to be spanned is short. However, longer distances present problems. There is a limit to how far apart the piers can be. If the distance between piers is too great, the beam may break under the combination of its own weight and the weight of the traffic it carries. Given that fact, you may be surprised to learn that some of the world's longest bridges are beam bridges. To make them safe, engineers use many piers placed close together to support the beam.

One of the most famous beam bridges in the world is Canada's Confederation (kuhn-feh-duh-RAY-shun) Bridge, which is also called the Northumberland Strait Bridge. The bridge, shown below, crosses the Northumberland Strait to join Prince Edward Island with Canada's mainland. The 8-mile-long bridge was completed in 1994. It is made of concrete and was designed to last 100 years.

The Confederation Bridge reaches across the narrowest part of the strait. You might expect that the engineers would design a very straight bridge to keep it as short as possible. However, the bridge actually has several slight curves. Experts believe this helps keep drivers alert and reduces the number of accidents.

The Confederation Bridge has a hollow channel inside it. Electrical and telephone wires run through this channel from the mainland to Prince Edward Island.

The Confederation Bridge has 65 piers, with 64 spans linking them. What is the average length, in feet, of a span? To find the answer, you must first multiply the number of feet in a mile (5,280) by the length of the bridge in miles (8). Then divide this figure by the number of spans (64) to find the average length of a span.

```
  5,280 feet per mile              660
x     8 miles             64 ) 42,240
 42,240 feet                  - 38 4
                                 3 84
                               - 3 84
                                   00
```

The average length of a span is 660 feet.

Beams were also used in a famous Scottish bridge built between 1883 and 1890. The bridge, shown on page 17, spans the **estuary**, or firth, of the Forth River and is known as the Firth of Forth Bridge. The long beams of the Firth of Forth Bridge, which project from 3 main towers, are not supported by a series of piers, as we saw in the Confederation Bridge. Rather, they are supported by **diagonal** tubes that project from the tops and bottoms of the towers. This type of beam bridge is known as a **cantilever** bridge.

The Firth of Forth Bridge was the first bridge built mostly of steel. About 54,000 tons of steel, 21,000 tons of cement, and 23,740 cubic yards of granite were used in its construction.

Suspension Bridges

In the late 1800s, many engineers began to favor suspension bridges. Suspension bridges get their name from the fact that the deck is suspended from huge cables that rest on top of tall towers. The cables carry the weight of the bridge to **anchorages**, which are buried in solid rock or concrete at the ends of the bridge. Suspension bridges are light and strong, and can easily span distances of more than 1 mile. Many people consider them more beautiful than other types of bridges. They often look light and graceful in spite of their great strength. However, they are also the most expensive type of bridge to build.

One of the most famous suspension bridges is also one of the first to be built. The Brooklyn Bridge was built between 1869 and 1883. It was the first suspension bridge to use steel for its cables. At the time of its completion, the Brooklyn Bridge was the longest suspension bridge in the world. From one anchorage to the other, the bridge is over 3,400 feet long. The deck is 85 feet wide and contains 6 traffic lanes.

More than a century after its completion, the Brooklyn Bridge is the second busiest bridge in New York City. Over 140,000 cars and trucks cross the bridge every day.

The deck of the Brooklyn Bridge is supported by 4 cables. Each cable is about .68 mile long and is made up of 5,434 wires. How many miles of wire were needed for all 4 cables? Solving this problem requires 2 steps.

First, multiply the number of wires in 1 cable (5,434) by the length of the cable (.68 mile) to get the number of miles of wire in 1 cable.

$$\begin{array}{r} 5{,}434 \text{ wires in 1 cable} \\ \times\ .68 \text{ mile} \\ \hline 434\ 72 \\ +\ 3\ 260\ 4 \\ \hline 3{,}695.12 \text{ miles of wire in 1 cable} \end{array}$$

Next, multiply the number of miles of wire in 1 cable (3,695.12) by the number of cables (4) to get the number of miles of wire in 4 cables.

$$\begin{array}{r} 3{,}695.12 \text{ miles of wire in 1 cable} \\ \times\ \quad 4 \text{ cables} \\ \hline 14{,}780.48 \text{ miles of wire in 4 cables} \end{array}$$

The total amount of wire needed for all 4 cables was about 14,780.48 miles. That's enough wire to reach more than halfway around the world!

The Brooklyn Bridge has 2 stone towers. Each tower contains about 42,580 cubic yards of **granite**. How many cubic yards of granite were used in both stone towers?

To find the total number of cubic yards, multiply the approximate number of cubic yards in 1 tower (42,580) by the number of towers (2).

$$\begin{array}{r} 42{,}580 \text{ cubic yards per tower} \\ \times\ \quad 2 \text{ towers} \\ \hline 85{,}160 \text{ cubic yards} \end{array}$$

A total of about 85,160 cubic yards of granite was used in the 2 towers.

How many cubic feet does this amount of cubic yards equal? First, find the number of cubic feet in a cubic yard. Since a yard has 3 feet, a cubic yard has 3 feet by 3 feet by 3 feet, or 27 cubic feet.

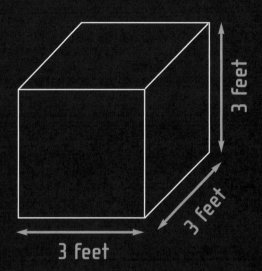

Next, multiply the number of cubic yards (85,160) by the number of cubic feet per cubic yard (27).

85,160 cubic yards
x   27 cubic feet per cubic yard
596 120
+ 1 703 20
2,299,320 cubic feet

A total of about 2,299,320 cubic feet of granite was used in the towers.

Another famous suspension bridge, built on the opposite side of the United States, is the Golden Gate Bridge. When the idea of building a bridge across the Golden Gate Strait was proposed in 1916, many people objected. Some said it was impossible. Others said it would cost too much money. Ferry operators who transported people across the strait feared they would be put out of work. Finally, however, people were convinced that a bridge was a good idea, and money was raised to build it. Construction began in 1933, and the bridge was completed in 1937.

Each anchorage is made of steel and concrete, and each weighs 60,000 tons. The steel used in each anchorage weighs 2,200 tons. How many tons of concrete were needed for both anchorages together? First, find out how many tons of concrete were used in each anchorage. Then, multiply that amount by 2 to find out how many tons were used in both anchorages.

Subtract the number of tons of steel (2,200) per anchorage from the total weight of 1 anchorage (60,000).

```
  60,000 tons total weight
-  2,200 tons of steel
  57,800 tons of concrete
```

Next, multiply the number of tons of concrete in 1 anchorage by 2.

```
   57,800 tons of concrete/anchorage
x       2 anchorages
  115,600 tons of concrete total
```

About 1,471 cubic yards of concrete were used for each $\frac{1}{10}$ mile of the road on the Golden Gate Bridge. The bridge is 1.7, or $\frac{17}{10}$, miles long. How many cubic yards of concrete were used altogether?

The longest suspension bridge in the world is the Akashi Kaikyo (ah-KAH-shee KY-kee-oh) Bridge, also known as the Akashi Strait Bridge. The bridge, completed in 1998, stretches 12,828 feet across Akashi Strait. It links the city of Kobe on the Japanese island of Honshu with the Japanese island of Awaji. The engineers faced particular challenges in designing this bridge. They had to create something that would be able to handle the powerful winds and earthquakes that often occur in Japan. The design they created can survive winds up to 180 miles per hour and earthquakes more powerful than the one that nearly destroyed San Francisco in 1906. Part of the design that makes the Akashi Kaikyo bridge so strong is a network of triangular supports beneath the roadway. This network is called a truss. The truss makes the bridge very rigid and at the same time allows high winds to pass right through the structure.

The main body of 1 of the anchorages contains about 183,400 cubic yards of concrete. The main body of the other anchorage contains about 13,100 more cubic yards of concrete than the first anchorage. How many cubic yards of concrete were used in the main bodies of both anchorages?

To find the number of cubic yards of concrete in the second anchorage, add 13,100 to the number of cubic yards in the first anchorage (183,400).

Next, add that figure to the number of cubic yards of concrete in the first anchorage.

```
  183,400 cubic yards
+  13,100 cubic yards
  196,500 cubic yards
```

```
  183,400 cubic yards
+ 196,500 cubic yards
  379,900 total cubic yards
```

There are about 379,900 cubic yards of concrete in the main bodies of both anchorages combined.

About 186,000 miles of wire were used in the cables of the Akashi Kaikyo Bridge. That's enough to go around the world 7.5 times! How much more wire was used in the Akashi Kaikyo Bridge cables than in the Brooklyn Bridge cables? You can use the information on page 20 to help you find the answer.

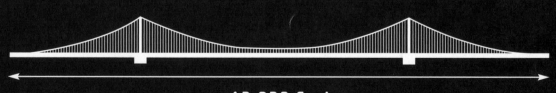

12,828 feet

A Famous Failure

One of the most well-known suspension bridges in the world is famous because it collapsed. The Tacoma Narrows Bridge, which opened on July 1, 1940, linked the mainland of Washington State with the state's Olympic Peninsula. As the table shows, the total length of the bridge and the size of its cables were about the same as those of the Brooklyn Bridge. However, the center span of the Tacoma Narrows Bridge was much longer than the center span of the Brooklyn Bridge.

	Tacoma Narrows Bridge	Brooklyn Bridge
total length	5,939 feet	6,016 feet
cable diameter	17.5 inches	15.75 inches
number of wires in 1 cable	6,308	5,434
length of center span	2,800 feet	1,595 feet

The long center span of the Tacoma Narrows Bridge often swayed back and forth in the wind, leading the people of the area to nickname the bridge "Galloping Gertie." On the morning of November 7, 1940—just 4 months after it was completed—wind speeds up to 45 miles per hour caused the bridge to sway and twist. Finally, at about 10:30 A.M., the bridge began to break apart and fall into the water below. The events were captured in the photographs shown on page 27.

How much longer was the center span of the Tacoma Narrows Bridge than the center span of the Brooklyn Bridge? How would you find your answer?

An earlier bridge across Tampa Bay collapsed after a large ship ran into it during a storm. To keep that from happening again, the new bridge's engineers placed large concrete islands, called dolphins, around the bridge's piers to protect them.

Cable-Stayed Bridges

Cable-stayed bridges look a lot like suspension bridges. Both kinds of bridges have tall towers and cables that support the roadway. However, in suspension bridges, the cables rest on the tops of the towers but are not attached to them. The cables are attached to the anchorages at the ends, and the anchorages bear the weight of the roadway. In cable-stayed bridges, the cables are attached to the tops of the towers, and the towers bear the weight of the roadway.

A famous cable-stayed bridge is the Sunshine Skyway Bridge, which crosses Tampa Bay and links the Florida cities of St. Petersburg and Bradenton. The bridge, completed in 1987, is 29,040 feet long—more than twice as long as the Akashi Kaikyo Bridge! The deck is built of about 300 concrete blocks that weigh on average about 220 tons each. About how many tons of concrete were used for the Sunshine Skyway Bridge's deck?

Multiply the average weight for a block (220 tons)
by the total number of blocks (300).

$$
\begin{array}{r}
220 \text{ tons per block} \\
\times\ 300 \text{ blocks} \\
\hline
000 \\
0\ 00 \\
+\ 66\ 0 \\
\hline
66{,}000 \text{ tons}
\end{array}
$$

Altogether, about 66,000 tons of concrete were used for the bridge's deck.
Can you figure out how many pounds that equals?
Remember, there are 2,000 pounds in 1 ton.

More Bridge Math

In this book, we've used math skills to learn something about a few of the world's famous bridges. Now that you know how to do it, you can use your math skills to learn something about some other famous bridges. Other

**Tower Bridge
London, England**

famous arch bridges include the Ponte Vecchio, or Old Bridge, in Florence, Italy, and the Rialto Bridge and the Bridge of Sighs in Venice, Italy. Other well-known suspension bridges include the Tower Bridge in London, England, and the Tsing Ma Bridge in Hong Kong. The Zakim Bunker Hill Bridge in Boston, Massachusetts, is another example of a cable-stayed bridge.

The engineers who design bridges use the same math skills we did in this book—and more—to solve dozens of problems so that they can make their bridges strong, safe, and beautiful. Perhaps someday you, too, will use your math skills to design bridges!

Glossary

abutment (uh-BUT-muhnt) The part of an arch bridge that carries the weight of an arch.

anchorage (ANG-kuh-rihj) In a suspension bridge, the part that anchors the cables and keeps them secure.

aqueduct (AH-kwuh-dukt) A man-made structure that carries water from its source to a place where it is needed.

cantilever (KAN-tuh-lee-vuhr) A beam or system of beams that extends from a tower and supports a span of one type of beam bridge.

cast iron (KAST I-uhrn) A type of metal made of iron and other elements that are melted together, then poured into a mold to create the shape needed.

deck (DEHK) The roadway of a bridge.

diagonal (dy-AA-guh-nuhl) Describing a line or figure that is positioned on a slant.

engineer (en-juh-NEER) A person who designs, constructs, or manages the construction of a structure, such as a bridge.

estuary (ESS-chuh-wear-ee) A water passage where a saltwater tide meets a freshwater river.

granite (GRA-nuht) A very hard rock commonly used in the construction of buildings.

horizontal (hor-uh-ZAHN-tuhl) Parallel to the horizon.

peninsula (puh-NIN-suh-luh) A portion of land surrounded on 3 sides by water.

pier (PEER) A structure made of concrete or metal used to support the ends of bridge spans.

span (SPAN) To extend across something. Also, the distance between supports on a bridge.

Index